# THE HUMAN BODY

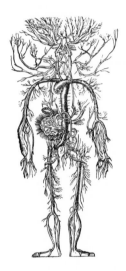 

*Two early engravings of human trees of veins and arteries. Treelike structures are an ideal way to maximize and connect points of contact within a finite space. We share many genes with trees, and early geneticists thought that the amount of DNA in an organism determined its prowess, until it was discovered that humans have the same amount as privet hedges, while lilies have 33 times more.*

First published in the United States of America in 2004 by
Walker Publishing Company, Inc.

Published simultaneously in Canada by Fitzhenry and Whiteside,
Markham, Ontario L3R 4T8

For information about permission to reproduce selections from
this book, write to Permissions, Walker & Company,
104 Fifth Avenue, New York, New York 10011.

Library of Congress Cataloging-in-Publication Data available
upon request
ISBN: 0-8027-1429-3

Visit Walker & Company's Web site at www.walkerbooks.com

Printed in the United States of America

2 4 6 8 10 9 7 5 3 1

# THE HUMAN
# BODY

### a basic guide to the way you fit together

## *Moff Betts, M.D.*

*with additional illustrations by*
*Borthwick, Ede, Huson, and Tweed*

Walker & Company
New York

*To Ruby, Eva, Hana B, Katie, Jessie, and Jack*

*Many thanks to the following for their illustrations:*
*Cecily Kate Borthwick (13, 15, 29, & 41), Caroline Ede (below, 3, 17, & 45),*
*David Goodsell (43t), Simon Huson (9, 19, 24, 26, 33, 37, & 39),*
*Miranda Lundy (5, 23, & 51) and Matt Tweed (11, 12, 21, 35b, & 47).*
*Others from Albinus, Vesalius, Winkles, and various Victorian engravers.*
*Terms italicized in the main text refer to the glossary on pages 52-53.*
*To learn more about less of the human body, buy a bigger book.*

*I'd like to thank: My Parents, Jeanne Jeffares, Justin Avery, Tony Prescott,*
*Adrian Finter, Guy Lederman, Brian Davies, John Deutsch, Tom Betts,*
*and Susy Tucker, for their excellence, vision, and humor.*

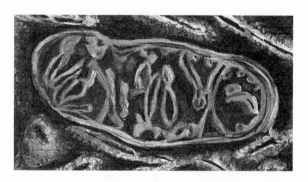

*A mitochondrion (magnified x 30,000), a descendant of ancient bacteria. Up to several thousand live in each cell in mitochondria-rich tissues like muscle and liver. The oxygen from the air you breathe goes to these powergrubs, who make your energy, and the carbon dioxide you breathe out. Their genes are in one circle of DNA, and are passed down only in your maternal and grandmaternal line, and so on, back to the first human mothers and beyond.*

# CONTENTS

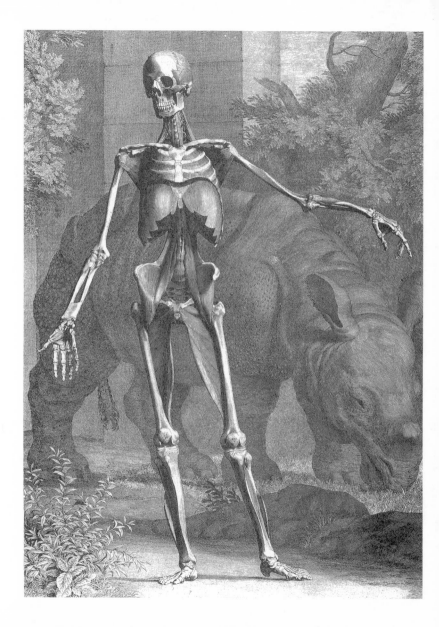

# INTRODUCTION

The human mind is so inquisitive, and our bodies are so intricate, that curious men and women have made our own flesh the most densely studied field of all. So when I was asked to write a tiny book about the human body, I protested that the topic was too large and complex to fit into such a limited space, and that the task was clearly impossible. As it turned out, the book had ideas of its own, and with the combination of cram and scatter used in these pages, you may find that your understanding deepens on a second read through.

What follows starts with some history of our investigations into the body, then proceeds to examine the microscopic mechanisms that have become the cornerstone of modern biology. The bulk of the book comprises a look at the functions of the major bodily systems, before ending with some examples of integration. If you are unfamiliar with the life sciences, the glossary of italicized terms beginning on page 52 may be invaluable.

Being human isn't all about constant awareness of our bodies; indeed some of our best experiences happen when we manage to forget about them altogether. Unique, doomed to death, and your only true possession, your body's frantic devotion to keeping you alive is roundly ignored most of the time, allowing you to get on with the brilliant business of life unfettered. But I hope that these few glimpses of our discoveries about the human body, many of them truly amazing, will leave you with the feeling of gratitude and awe that this enigmatic and unparalleled form deserves.

*Franksbridge Manse, 2004*

# Round Windows
## *see for yourself*

---

Little was known of the interior layout, let alone the workings of the body, until remarkably recently. As Europe floundered in the Dark Ages, when ideas like drinking pig dung for pleurisy were in vogue, cosmopolitan Baghdad and cerebral Teheran shone as centers of science, while India and China still had sophisticated systems from antiquity. Tellingly, dissection was fairly taboo in all cultures until the Italian Renaissance, where the anatomists who made the first systematic cuts named their discoveries, like Fallopian and Eustachian tubes, after themselves. Animal-based errors and blind conjecture, fabricated by Galen in second-century Alexandria, ruled in the West until Vesalius published the first definitive map of the body in 1543, informed by an ample supply of executed convicts.

Mankind's ensuing autopsy ("see yourself") has exploded us into innumerable bits, mostly studied in dead humans or half-dead furry mammals, a far cry from a living whole.

The clearest piece of living tissue you can see without cutting into someone is an iris; looking at someone else's avoids the left-right flip in the mirror. No two are the same, not even in one head. *Irismeisters* can see your whole body mapped onto the iris, which is mainly muscle colored by opaque pigments shielding the light-sensitive retina at the back of the eye from overexposure. Outer radial muscles dilate the pupil in the dark, balanced by an inner circle that constricts in the light, increasing depth of field. Translucent *aqueous humor*, filtered from blood by tendrils behind the iris, flows freely in its catacombs, even during rapid motion.

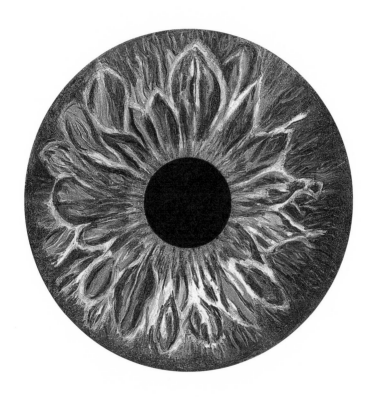

# ASIAN FUSION
## *ancient Indian and Chinese systems*

The basic premise of all ancient systems is that we are part of a larger living body, the Cosmos, and are shaped by the same forces.

In Chinese, unity divides into *yin* and *yang*, which revolve as a vortex of *chi*, energy that flows through all things.

Chi's flux through humans has long been charted, by observers who presumably had an idea of the likely internal apparatus from butchery and imagination. Meridians are the chi-flux channels they sensed on the body, which map onto 12 internal organs, with central connections out of sight. They guide the flow of organ-specific chi around your innards and to your extremities.

Every one of the 361 vortices used as access points along the meridians has its own quality of chi, reflecting the state of its related organ, itself ruled by the interplay of 5 elements *(p.58)*.

India's 2 middle meridians form a double spiral that snakes up the spine, channeling *kundalini*. This is rising *prana* (chi), whose yin and yang are *shiva* and *shakti*. Where the snakes cross, 7 midline vortices form *chakras* (*p.45*), spinning, colored prana funnels. What makes you alive is the ability to accept, transform, and radiate chi, in all its infinite guises, back to the universe.

As well as mapping inner activities onto the surface, perennial systems also encompass the various influences of human interfaces with nature, such as diet, habits, and locale, so your total level of integration with the Cosmos determines your vitality.

Western science interprets prana as electricity, central nervous *gating* effects, explaining the ability of acupuncture to anesthetize.

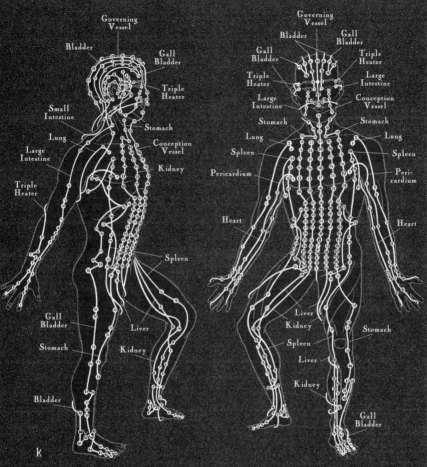

Each meridian is a pathway of acupuncture points forming a circuit with an organ via inner pathways. The 12 bilateral meridians then also form a continual circuit in the order lung, large intestine, stomach, spleen, heart, small intestine, bladder, Kidney, pericardium, triple heater, gallbladder, and liver, connected through additional internal meridians. The 2 central meridians bring the total to fourteen.

# ELEMENTS AND HUMORS
*man mirrors nature*

Medieval folk saw the body as composed of the same number of elements as nature, which was usually 3, 4, or 5.

Inner versions of fire, air, earth, and water swirled around as humors, blending in the body, and painting its faces, seasons, and qualities. Like the weather, humors had mild, stormy, heavy, and beautiful days inside each person. As climatic patterns sculpt local landscape, humors revealed the deeper colors of your complexion.

Humors' carrier fluids were only visible in the raw during an illness—the body's brave attempt to restore humoral harmony. Too much humor (in the wrong place) could be cured by efflux of blood, phlegm, pus, or bile, or any combination thereof.

The Swiss mystic and iconoclast *Paracelsus* saw the ethereal humors being mirrored by physical counterparts, basing his system on an alchemical trinity of salt, sulfur, and mercury. As in nature, these combined inside the body, making fixed, mutable, and cardinal compounds, which conferred structure, energy, and communication, like the *molecules* now found in microbiology.

Subtle blending effects can also be seen in modern *endocrinology*, the character of whose *hormones* emerges starkly in various syndromes of excess or depletion, giving clues to their patterns in health.

Four chemical elements (H, N, C, and O) combine in the 4 *bases* of *DNA*, the molecule that carries hereditary complexion. A fifth element, phosphorus, joins the pentagonal sugars of DNA's backbone, to make a long chain with a mirror image, the 2 strands resolving as the serpentine double helix (*p.23*).

hot

**CHOLERIC**
nervous - plasma - fire
hydrogen

**SANGUINE**
arterial - blood - air
nitrogen

dry

the

**HUMOURS,**

tempers,

elements,

&

other esoterica

phosphorus

wet

**MELANCHOLIC**
digestive - bile - earth
carbon

cold

**PHLEGMATIC**
lymphatic - mucus - water
oxygen

7

# PRE-CONCEPTION
*two generation gaps*

The egg bobbed at anchor, with a 1,000 others, made in your mother's ovary before her own birth. A generation later it set sail and met your father's sperm swimming into port, spawned just two moons earlier, as one of that day's batch of 300 million.

Apart from the fusion of *gametes* (sperm and egg) at the moment of conception, new cells are made by cell division. Most cells arise from *mitosis* (*p.14*), but gametes are created by *meiosis*. A gamete has 23 *chromosomes*, the long strands of DNA that carry *genes*, so when 2 gametes unite, the *conceptus* has 46, which are copied into every cell in your body. The chromosomes in the gametes that made your first cell were remixes of your grandparents' chromosomes, made in prophase I of meiosis, when each of your parents' parents' pairs swapped parts of their limbs. So you are a meeting of your parents, but a proper mix of your grandparents.

Girls' 23rd chromosomes are Xs from both parents, but boys inherit their father's father's Y instead of their mother's X. Girls have 2 Xs, but inactivate one in every cell to form a *Barr body*.

Vanguard sperm batter the giant *ovum's* shield before one straggler gets in, bringing its 23 chromosomes to join the 23 that the ovum doesn't discard at this point (she waited for ages in late anaphase II), making a unique new set of 46 chromosomes.

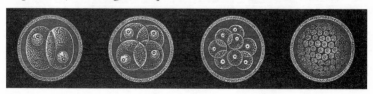

8

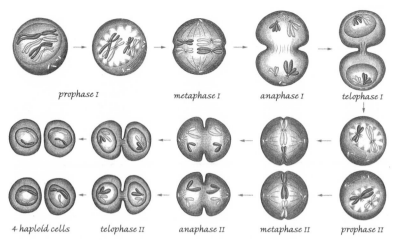

*prophase I*      *metaphase I*      *anaphase I*      *telophase I*

*4 haploid cells*      *telophase II*      *anaphase II*      *metaphase II*      *prophase II*

*MEIOSIS: Unique new eggs and sperm are created from parental DNA.*

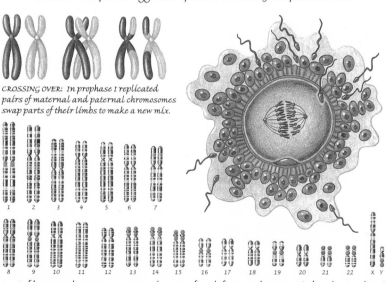

*CROSSING OVER: In prophase I replicated pairs of maternal and paternal chromosomes swap parts of their limbs to make a new mix.*

1 2 3 4 5 6 7

8 9 10 11 12 13 14 15 16 17 18 19 20 21 22 X Y

*A set of human chromosomes, 23 pairs, one of each from each parent (a boy, last pair XY).*

# AND DNA
## *cooking up shapes in the protein kitchen*

Down in molecule sea, DNA functions as a cookbook. Written in a 4-base "alphabet", *a c g t*, humans' 6-billion–base *genome* is 97% rhythmic, being long *tata*-rich passages. Against this babble the odd coherent recipe for a "shape" pops up, as a readable string of 3-letter *codons*, like *tac-gtc-gag-cct-cac-gtt-cat-ctc-taa-ttg-ccc-atc*.

The recipes are for chains of *amino acids*, which form *proteins*, the next level of molecular shape and function. DNA also holds recipes for *RNA*, her busy single-stranded, single-use sister, who runs the kitchen and makes all the utensils. She copies recipes, chops and peels, and builds multiprotein dishes, while dreamy DNA just lets herself be read, replicated, and rarely modified.

DNA's left- and right-hand helixes are mirror images, since the bases pair off, *a-t* and *c-g*, linked by *hydrogen bonds* across the spiral innerspace. RNA reads the counterclockwise thread; the back-up side encodes repair and lets the twin zipper *supercoil* into a superhelix. A gene is a set of recipes coding for a finished protein (or RNA), and "switches on" when all segments uncoil, unzip, and are read.

Most of the time DNA lies partially coiled inside the *nucleus*, only assuming the familiar X shape during cell division, when two copies of a replicated chromosome are still joined at the hip.

Knowing how short one life is, DNA hops across chromosomes, species, and generations. *Viruses* hijack your kitchen, while *bacteria* pass on their crazy new recipes, helping evolution along. New cookbooks were compiled long ago by pioneers like seasquirts and lungfish, to whom we owe much of our structure.

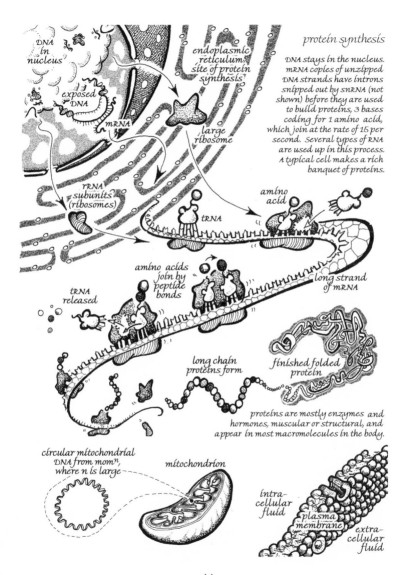

*protein synthesis*

DNA in nucleus

endoplasmic reticulum, site of protein synthesis

exposed DNA

mRNA

large ribosome

DNA stays in the nucleus. mRNA copies of unzipped DNA strands have introns snipped out by snRNA (not shown) before they are used to build proteins, 3 bases coding for 1 amino acid, which join at the rate of 15 per second. Several types of RNA are used up in this process. A typical cell makes a rich banquet of proteins.

rRNA subunits (ribosomes)

amino acid

tRNA

amino acids join by peptide bonds

long strand of mRNA

tRNA released

long chain proteins form

finished folded protein

proteins are mostly enzymes and hormones, muscular or structural, and appear in most macromolecules in the body.

circular mitochondrial DNA from mom$^n$, where n is large

mitochondrion

intra-cellular fluid

plasma membrane

extra-cellular fluid

# THE EARLIEST SURVIVOR
## *one stormy night about four billion years ago*

Organic life sprang from one first cell formed amid a swarm of molecular mosaics in a long-lost pool. Just how all the ingredients coalesced in the right circumstances, nobody knows, but happily, the mother of us all survived, replicated, diversified, and evolved.

After the dark early eons, cells used light, carbon dioxide ($CO_2$), and water to make carbohydrates and oxygen, as plants still do. With atmospheric oxygen rising wildly, one cell mastered a chain reaction that used this fiery gas to turn carbohydrates and water into $CO_2$ and energy, as we still do. This genius made it welcome inside other cells, thus furthering the tradition of *endosymbiosis*. Hundreds of descendants of these *archaebacteria* live inside each of your cells, as *mitochondria*, and the air you breathe is for them.

Promitochondria weren't the only cohabitants in early cells but have kept their genes and reproduction largely to themselves. *Eukaryotic* cells, which make up complex organisms like plants and humans, were made by several organisms, with pooled genes evolving into a nucleus. So our smallest living subunit, the cell, itself began as a cooperative of even smaller earlier organisms.

Prokaryotes (bacteria) retain the ability to exchange genes (*below*), as easily as we swap ideas, the global bacterial superorganism accessing a common gene pool in order to adapt to local niches.

12

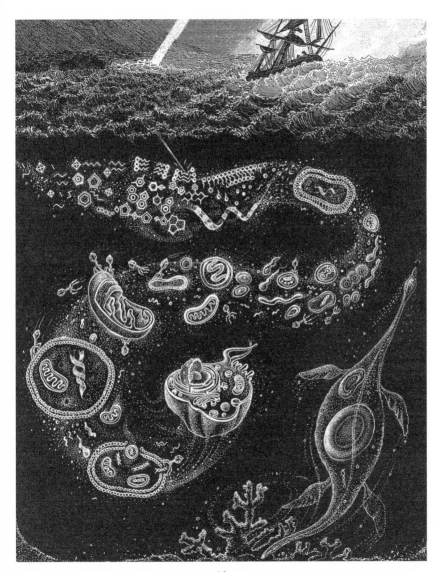

# THE CELL
## *life's electric microcosm*

Each of your trillions of cells differentiates into one of about 300 specialized types by switching on only a few percent of the genes in its nucleus. This gives rise to specific proteins, which engineer other proteins, fats, and sugars to build the cell's structures, which define its function. Proteins' 3-D diversity, helped by some fats, lets them recognize and transform other molecules, act as specific signals, and do flashy tricks like sense the wavelength of light.

Cells' internal soup differs radically from the extracellular fluid, operations taking place inside the thin fluid *plasma membrane*. This is studded with protein-based structures, mainly *channels*, governing what enters and leaves, *receptors*, transducing signals from other cells, or the ubiquitous *Na-K-ATP pumps*, exchanging sodium (out) for potassium (in). The ion fluxes that are thus set up mean that cells are electrified, negative inside and positive outside. This is crucial for all cells, not just muscles and nerves, for import and export processes, and the optimization of *enzyme* function.

Na-K-ATP pumps use half your total energy, which is basically your available *ATP*, made by mitochondria with a proton–electron transfer chain that uses food and oxygen to fuel ATP synthesis.

Cell division, mitosis (*below*), happens at intervals from every few hours in high-turnover tissues like front-line white cells (*p.34*), to probably never again, in the case of most of your brain cells.

14

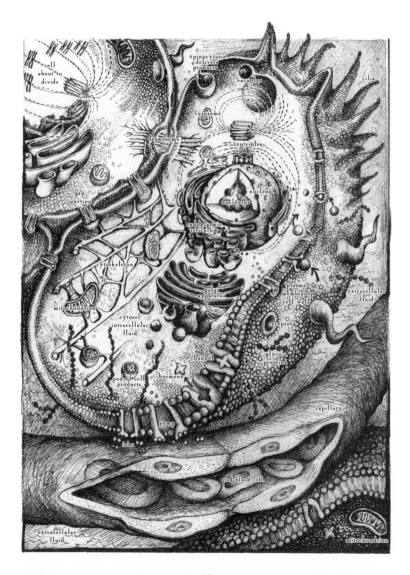

cell about to divide

cilia

pinocytosis to deliver products

vacuole

lysosome

desmosome

centrioles

gap junction

nucleus

nucleolus

endoplasmic reticulum

cytoskeleton

plasma membrane
cell wall

extracellular fluid

mitochondrion

golgi apparatus

cytosol intracellular fluid

receptor

glyco proteins

channel

pods of cell products

hormone

capillary

pump

ions

red blood cell

extracellular fluid

mitochondrion

15

# TISSUES
## *saltwater archipelagos*

Tissues are communes of similar cells performing the same tasks; several make up any organ. Tissues fall into 4 classes.

Epithelial tissues are highly selective about what they allow to pass through. They form linings of hollow and tubular organs, secreting mucus to protect and extend the border they control.

Connective tissue includes wandering and static cells, which deposit artifacts outside their walls to build infrastructures and general meshwork, as well as define your immune boundaries.

Muscle tissue comes in 3 types: Skeletal muscle contracts linearly under conscious control; smooth muscle squeezes circularly, narrowing the tubes in whose walls it writhes, subconsciously guided by *autonomic* nerves; and heart muscle's tireless cells also pass on electricity to their neighbors, at differing speeds in order to synchronize the pattern of cardiac contraction.

Apart from a few nerve cells in the gut, which have a mind of their own, all nerve tissues are linked in one giant electrical system.

Like Gaia, and life in general, we are two-thirds saltwater, which bathes and joins all tissues. The 3 primary fluids are: intracellular fluid, which varies according to cell; extracellular fluid, the tidal sea that bathes cells; and blood (*below*), half cells carrying gases and half fluid, *plasma*, carrying dissolved minerals and molecules large and small between the different tissues.

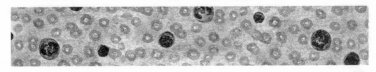

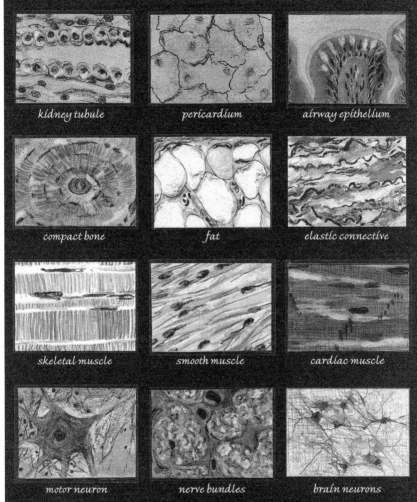

kidney tubule

pericardium

airway epithelium

compact bone

fat

elastic connective

skeletal muscle

smooth muscle

cardiac muscle

motor neuron

nerve bundles

brain neurons

Three examples of each of the 4 major classes of tissue:
epithelial, connective, muscular, and nervous.

All magnifications are about 400x, except nerve bundles (1000x).

# EMBRYOLOGY
## *one cell becomes human*

*Uterine milk* nourishes the embryo during the rapid cell division of the first few days. On day 5 the bundle of 32 cells bores a hole in its coat and implants into the wall of the womb, which feeds it, helped by the *yolk sac*, until the *placenta*, made jointly by mother and baby, is fully functional by the end of the first month.

Three primordial "germ" layers (*below*) develop into all tissues: *ectoderm* forms brain, nerves, and outer skin*; mesoderm* makes solid organs, flesh, and bone; *endoderm* forms most epithelial tissues.

The mysterious influences that make embryonic cells specialize and grow in patterns are still secret, but guidance seems to come through prenatal master genes, local signal gradients near the developing blood vessel tree, and pilot cells that die before birth.

The busy infolding, migration, and reunion of diverse pockets, arches, and tubes of cells proceeds so quickly that by 2 months it is recognizably human—and has made nearly all its little organs except the *gonads*. In boys, a signal cascade in week 7 is started by the *Sry-1* gene on the tiny Y chromosome, which acts on some cells that crawled from the yolk sac to the lower pole of the developing kidney, resulting in male genitals, with all their hormonal consequences. In girls, X–driven maturation continues, undiverted, to form womankind.

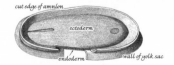

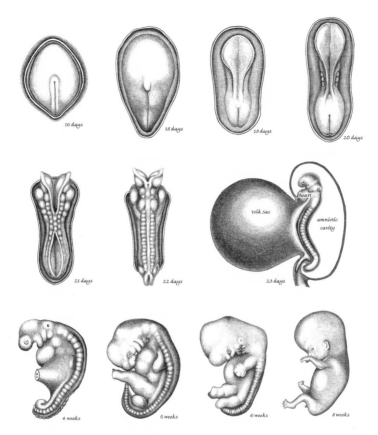

*16 days*  *18 days*  *19 days*  *20 days*

*21 days*  *22 days*  *23 days*  yolk sac  heart  amniotic cavity

*4 weeks*  *5 weeks*  *6 weeks*  *8 weeks*

*Early cells are totipotent, meaning they can become any type of cell, but gradually lose this ability as they differentiate into germ layers, then tissues and organs. A midline stripe, the early nervous system, first gives the embryo a head and tail, and the fastest-developing early organs are the brain and heart. Five arches migrate from the top of the heart to form head and neck structures, like ears, cranial nerves, and the thyroid, the earliest endocrine gland. Buds from the primitive gut form the lungs and liver, which begins making the first blood cells. The lower body develops from 44 paired segments, which give rise to bones, muscles, nerves, and inner organs. Limb buds are forming by 7 weeks, as is the young face. After 8 weeks, growth rather than differentiation takes over.*

# THE DISSECTOR'S DREAM
## *systems and organs*

Three easy things you can say about the body are, it was once one cell, it's alive, and it is an organized, integrated whole. The subunits the body gets divided into, be they cells, tissues, organs, or systems, are defined by specialization of form and function.

At a simple level, the body runs a set of cooperative systems, each performing a major task, like blood circulation. Any given system is made up of contributions from several organs, which themselves can be localized collections of tissues, like the heart, kidneys, or intestines, or diffuse, like blood vessels or nerve trees.

Any tissue contains a specific array of diverse cells, whose vast range of specialization makes it possible for huge beings like us to exist. Bathing in the warmed and highly modified sea we have trapped inside the dead matter of our waterproof skin, cells function slavishly, told what to do by signals from elsewhere.

The brain senses and controls the activity of the whole body, sending signals as short-term nerve impulses or longer-term control hormones (*p. 44*). The brain can override any other processes trying to happen, especially in adults, or, for example, Tibetan monks who can learn how to sweat in the snow naked.

The cartoons depicted opposite mostly have their own pages in this book, but one cannot consider them in isolation, since all perform functions that are crucial to the operation of the others. The body's systems are completely interdependent, and all intercommunicate, but quite how you are able to imagine the taste of your favorite food isn't yet known.

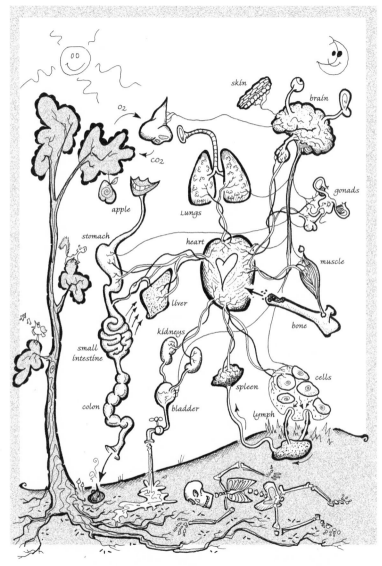

# FOUR PLUS ONE
## *is phive*

Variations in human shape are started off by your genes, and the 3 basic body types seem to relate to the dominance of one of the 3 embryonic germ-cell layers and their tissue densities: ecto- (wiry), endo- (rounded), and meso-morphs (muscular) (*p.57*).

However, there are many constant features to our frames. The lengths of the gullet and intestinal segments, for example, vary very little between tall and short people. Phi, $\phi$, the golden section, which occurs in organic growth and pentagrams, defines the change of your center during growth. A baby's midpoint is his or her belly button (the past), with the genitals a golden section away from the crown; but by adult life, the genitals (the future) are central, with the belly button $\phi$ from the feet. The lengths of the bones of the arm and hand also display a golden-section series, and $\phi$ occurs in the geometry of DNA (*opposite*).

Humans, like DNA, show 4-plus-1 patterns, as 4 limbs and a head protrude from the torso, and hands sprout 4 fingers and a thumb. Like fingers and toes, children have 5 teeth in each quadrant of their mouths; 8 more per quadrant develop and emerge later. The total amount of teeth you chew your way through in one life is thus 20+32, a full deck of 52.

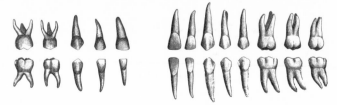

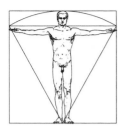

*Span is equal to height in an average adult, with the gonads halfway up or down. In a young child the belly button is in the middle. The golden section, φ, marks the position of the belly button in adults and the gonads in children.*

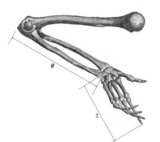
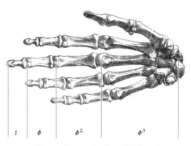

*The primary bending places from fingertip to elbow relate to their neighbors by the golden section, here 1.618, found in pentagrams and the water molecule.*

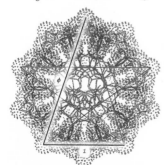
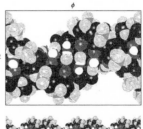

*A DNA cross-section shows the 10-steps-per-spiral nature of the molecule; a single twist of DNA fits a golden-section rectangle.*

# EARTH AND FIRE
*transformation of food in gut and liver*

The pleasant thought of food, which makes you go and find some, also stimulates the secretion of digestive enzymes from the pancreas, and bile from the gallbladder. These break down food into microscopic morsels in the small intestine, whose *villi* (*below, left*) absorb them over a surface area the size of a tennis court.

Nutrients are intelligently transported into villus cells and enter the blood through *capillaries*, which feed this rich soup, via the portal vein to the liver, where it mixes with fresh arterial blood in the porous cells of liver lobules (*below, right*). Molecular wizardry transforms whatever arrives into whatever you need—for instance, making sugars out of proteins—and renders all sorts of poisons harmless or usable. Most nutrient distribution proceeds from the well-named liver. Prometheus, chained to the rock of earthly life, had his pecked out every day by a sea eagle. Every night it grew back to full size. Your liver sacrifices its resources and integrity to meet the bodily demands of each day, and by night it restores order, mends itself, synthesizes new supplies, and makes ready for the sea eagle that comes with the sunrise.

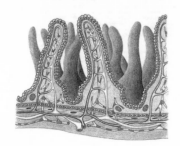 

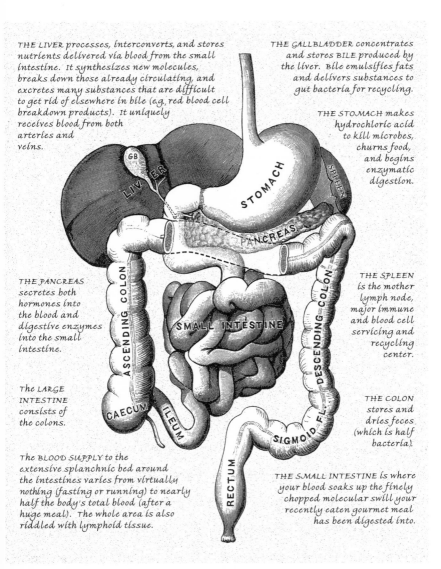

THE LIVER processes, interconverts, and stores nutrients delivered via blood from the small intestine. It synthesizes new molecules, breaks down those already circulating, and excretes many substances that are difficult to get rid of elsewhere in bile (e.g., red blood cell breakdown products). It uniquely receives blood from both arteries and veins.

THE GALLBLADDER concentrates and stores BILE produced by the liver. Bile emulsifies fats and delivers substances to gut bacteria for recycling.

THE STOMACH makes hydrochloric acid to kill microbes, churns food, and begins enzymatic digestion.

THE PANCREAS secretes both hormones into the blood and digestive enzymes into the small intestine.

THE SPLEEN is the mother lymph node, major immune and blood cell servicing and recycling center.

The LARGE INTESTINE consists of the colons.

THE COLON stores and dries feces (which is half bacteria).

The BLOOD SUPPLY to the extensive splanchnic bed around the intestines varies from virtually nothing (fasting or running) to nearly half the body's total blood (after a huge meal). The whole area is also riddled with lymphoid tissue.

THE SMALL INTESTINE is where your blood soaks up the finely chopped molecular swill your recently eaten gourmet meal has been digested into.

GB

LIVER

STOMACH

SPLEEN

PANCREAS

ASCENDING COLON

SMALL INTESTINE

DESCENDING COLON

CAECUM

ILEUM

SIGMOID FL.

RECTUM

# AIR AND WATER
## *lungs mirror kidneys*

In and out endlessly like the tides, breath constantly refreshes the blood, which supplies oxygen to mitochondria, and removes the $CO_2$ they synthesize. Blood–gas levels are monitored by the *brainstem*, which adjusts the rhythm and depth of inspiration.

*Bronchi* branch about 25 times, ending in bundles of elastic sacs named alveoli (*below, left*), which have a villi–sized total area (*p.24*). Over this, inhaled air exchanges gases with *hemoglobin* inside red blood cells, flowing at twice the usual speed, in the lung capillaries.

Kidneys' *glomeruli* (*below, right*) have some structures also found in alveoli. While the lungs expand to deal with gas balance, the kidneys head inward to regulate fluids. Kidneys take as much blood as the brain, and send signals that regulate blood pressure and make you thirsty. They filter the blood, then reabsorb 99% of the filtrate, to equalize salts, acids, volume, and composition.

Of the lungs' blood supply, a Syrian *Qanun* (*c.*1250) says

> *"When the blood has become thin, it is passed into the artery to the lung, to mix with the air. The fine parts of the lung then sieve into its vein, which arrives at the left of the two cavities of the heart."*

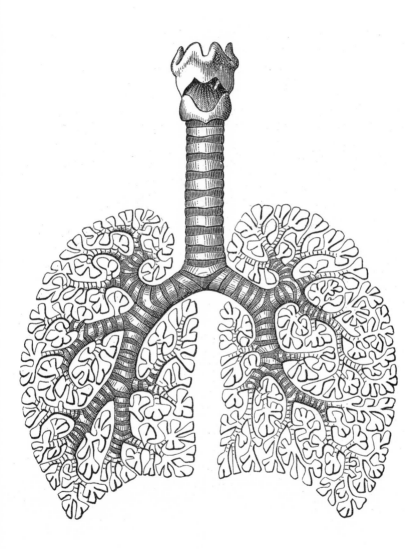

# THE HEART
*two beat as one*

---

The heart starts life as 2 primitive tubes that unite to form a cylinder, which is beating by the end of the embryo's third week. It grows, twists, and segments into a 4-chambered pump that *pulses* nonstop about a 100,000 times a day.

Each heartbeat is an electrically fired synchronized contraction, which initiates in a node of *pacemaker cells* in the right *atrium*. The pulse is differentially conducted over the heart to become a rising spiral wave. The easily detected electromagnetic field of this activity is often shown in 2-D as ⌁. In each beat, the 2 atria squeeze pooled blood through the *atrio-ventricular* valves into the *ventricles*, which wring it out in a vortex through the two *semilunar* valves. The lub–dub sounds are the 4 valves shutting, in 2 pairs.

This twin pump contracts as one, but runs a pair of circuits. The right heart collects blood from the body and sends it to the lungs, from where it returns, freshly aired, to the left heart, which pumps it around your body again, including to itself, via the frugally narrow coronary vessels. Half of all *blood flow* at any time goes through your lungs, and at the middle of this endless double loop is the heart, ceaselessly circulating the blood to aerate the body.

Cardiac cells are able not only to contract but also generate, transmit, and coordinate electricity, like a muscle-nerve hybrid. The electromagnetic field of each beat pulsing through the brain's sensitive nerves may be why the old ones say the heart is the deep seat of the mind. The *cockles* of your heart are warmth-sensitive organelles whose anatomical location has yet to be discovered.

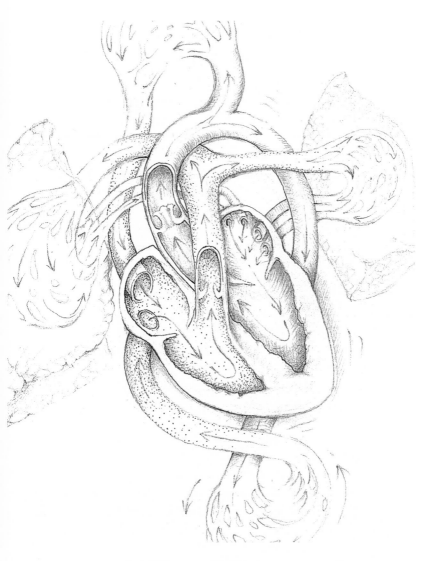

# VEINS AND ARTERIES

## *two trees unite*

The fact that the blood circulates endlessly was not discovered until 1628 by the no-nonsense Englishman William Harvey.

The arterial tree from the left heart branches to supply the body with blood, carrying oxygen, food, and signals. Between the blood and tissues is the micromesh of capillaries, tiny tubes that a red cell can just squeeze through, where gases, molecules, and ions soak out through the thin capillary walls. Here the blood picks up cell produce and waste, including $CO_2$, before a twin tree converges into veins, returning old blue blood to the right heart.

Blood cells are bred in the bone *marrow*. Red blood cells have no nucleus, so are disposable disks that maximize surface area for gas exchange. The molecule that carries gases inside red cells, and gives blood its color, is hemoglobin, whose 4 protein-pigment subunits each contain 1 iron atom. Each subunit carries 1 $O_2$ or $CO_2$ molecule under armed guard, swapping gases, one for the other, according to local conditions in the tissues and lungs.

Red cells are loaded with *ATP*-generating systems that don't use oxygen and have no internal structures. Unlike other cells, their outer surface is electronegative, attracting them to the positive extracellular fluid space, which they serve. So an electric current is flowing in your vessels, with the heart as the battery. Artery-vein pairs often run with a nerve that shares their name.

A red blood cell lasts about 4 months before it is chomped by a *macrophage*, in the spleen or liver, at the same rate as your marrow makes new red cells, which is about 3 million per second.

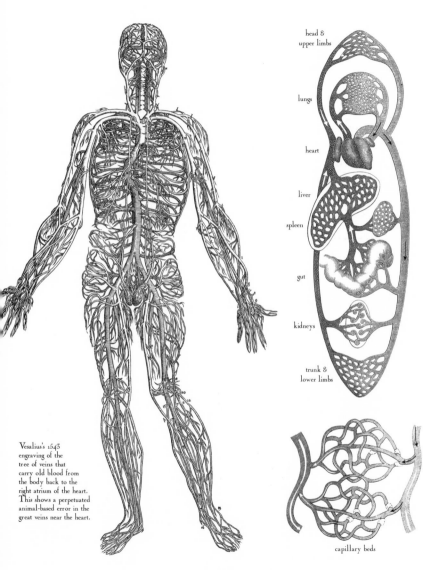

head &
upper limbs

lungs

heart

liver

spleen

gut

kidneys

trunk &
lower limbs

Vesalius's 1543
engraving of the
tree of veins that
carry old blood from
the body back to the
right atrium of the heart.
This shows a perpetuated
animal-based error in the
great veins near the heart.

capillary beds

31

# LYMPH PATROL
*apprehension of miscreants*

Like the sea, extracellular fluid contains flotsam and jetsam, so you have a lymphatic tree to soak up the overflow and detritus from around cells, and screen it for undesirables. Pure, filtered lymph is finally delivered back to the great veins near the heart.

White blood cells are generated from the same stem cells as red cells, as are *platelets,* which plug any gaps in vessel walls. White cells run your immune system and don't stay in the blood but venture out, looking for trouble in the lymphoid network.

Where you touch the outer world is where grit and microbes get in, so gut, lung, and skin particularly teem with lymphoid tissues. Here, various sorts of white cells come for brief or long stays, and messages flurry between them anywhere they gather, in order to coordinate activity and alert each other to damaging intruders.

The most mobile of all cells is the *lymphocyte*, which squeezes through gaps it makes between capillary cells, gropes around in tissues, sampling forms *(p.35)*, then crawls into the blind-ended lymph ductules, to communicate its findings to other white cells.

Lymphocytes stop off and collaborate in nodes along this tree. Swollen lymph nodes (glands) are only doing their job, packed full of lymphocytes, macrophages, and other types of white cells. Macrophages evolved from *amoebae*, and can engulf and digest material. Some innately recognize dust, while others eat anything dubious looking, and show chewed-up bits of it to lymphocytes, whose verdict on their potential hostility will inform other macrophages to destroy and recycle whatever needs eliminating.

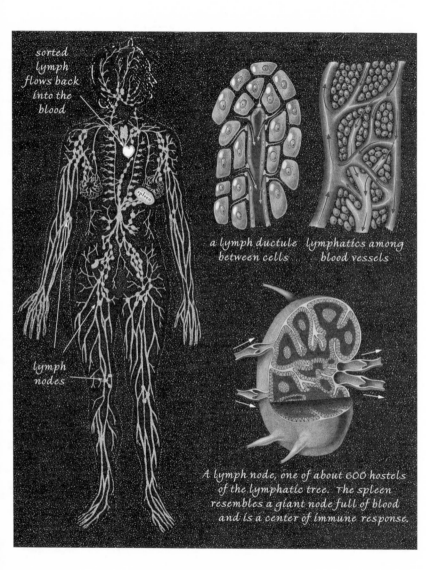

sorted
lymph
flows back
into the
blood

spleen

a lymph ductule    lymphatics among
between cells        blood vessels

lymph
nodes

A lymph node, one of about 600 hostels
of the lymphatic tree. The spleen
resembles a giant node full of blood
and is a center of immune response.

# IMMUNOLOGY
## *discrimination of self*

Childhood is when you learn the most, and nowhere more so than in the *thymus*, the gland that sits close to the heart. It spends your prenatal months killing any revolutionary lymphocytes that mutiny on sampling your own thymic cells, which is the majority. So, by the time you are born, your surviving lymphocyte army only reacts to unfamiliar forms that aren't recognized as you.

The main shapes that identify cells as yours are *MHC* proteins, and lymphocytes brandish other proteins that dock with these. If docking falters, the lymphocyte smells a rat and sends messages that amplify far and wide to bring the white-cell cavalry. To avoid whimsical wars being waged, several white cells agree to arouse one another's martial sides before letting full hostilities commence.

You have 2 basic weapons: custom-cloned *granulocytes* that recognize and kill certain microbes; and *antibodies,* bespoke proteins produced by lymphocytes in response to an *antigen*. Antibodies cooperate to form attack complexes or bind to microbes while signaling "eat me" to a macrophage. Few intruders are hostile, and many ride along helping out, or pretend to be you in sneaky ways.

Veteran white warriors remember their battles and can fashion weapons at short notice when the progeny of old enemies return. They are trained during childhood infections, when the immune army is learning military tactics. The thymus shrinks after your first birthday, and by dotage it has all but been replaced by fat cells. So as the years roll by, the school of discrimination between self and nonself gradually fades, a thymic idea of what one life is.

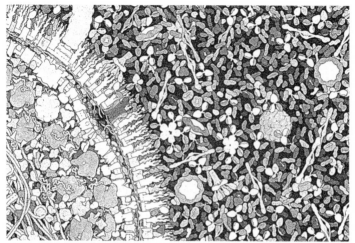

*Above: an IgG antibody-C1 complex binding to a flagellated E. coli bacterium (left), triggering a cascade that leads to an attack complex piercing the intruder's wall. Below: a happy lymphocyte samples you, while several cells cooperate to wage war.*

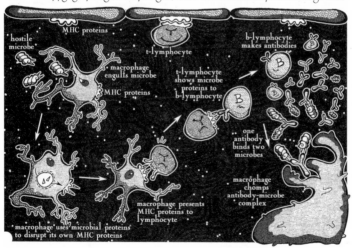

# NERVES
## *an electrical intranet*

---

The major divisions of the nervous system are *sensory*, *motor*, and central (the brain and spinal cord). Sensory *dendrites* bring data from the 5 peripheral senses into the central nervous system, where all the processing, integration, and thought take place.

Apart from a few *ganglia*, all nerve-cell bodies, *neurosomes*, live in the brain and spinal cord, where each enjoys the support of several *glial* cells. Motor nerve *axons* extend to supply muscles and are either "voluntary" to skeletal muscles, or *autonomic*, subconsciously controlling smooth muscles in all your tubes, glands, and organs.

The *voltage gradient* across nerve-cell membranes is similar to that at the center of a thunderstorm (*p. 14*), and it fluctuates both rhythmically and in response to incoming nerve impulses.

When the sum of local voltages reaches a threshold at the *axon hillock*, the cell fires, and a wave of temporary voltage-gradient reversal thrills along the axon, as successive *voltage-gated* channels open to let sodium ions flood in for a millisecond. This moves at walking pace in the slowest sensory nerves, and at 250 m.p.h. when it hops along gaps between sleeves of *myelin* around motor axons.

Axons terminate at *synapses*, where a *neurotransmitter* is released. This travels across the gap, usually to another nerve's dendrite or to a muscle, or sometimes to another axon. Here it makes the muscle contract (*p. 42*), or excites or *inhibits* the next nerve cell's frequency of firing. All nerve cells fire regularly, rate and rhythm defining their signal in the circuitry, and patterns of inhibition are what carve meaning in the white noise of the nervous system.

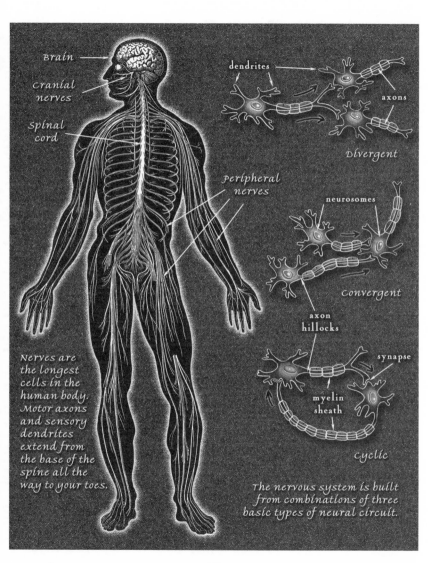

Brain

cranial nerves

Spinal cord

Peripheral nerves

Nerves are the longest cells in the human body. Motor axons and sensory dendrites extend from the base of the spine all the way to your toes.

dendrites

axons

Divergent

neurosomes

Convergent

axon hillocks

synapse

myelin sheath

Cyclic

The nervous system is built from combinations of three basic types of neural circuit.

# THE BRAIN
## *nobody's cracked the walnut*

Intelligence had to find a structure on which to discriminate itself, and it went for the brain. *Homo's* uniquely hefty walnut (only dolphins come close) localizes its subroutines so much that you even have a special area to hold images of your family, while keeping their names and maps of where they are in other zones.

Given the countless processes that make the electrical pattern of a single thought, it's astonishing that we're capable of unified consciousness. Awareness seems to flicker on a thin (3 mm) gray outer layer of *cortex*, which is all neurosomes, while the bulky *white matter* is all connections. The musical right hemisphere and logical left converge on the midbrain, where conscious thought merges with subconscious control and integration processes, while your deepest feelings and memories revolve around the *limbic system*. Most nerves swap sides through the brainstem, so the left brain corresponds to the right body, and vice versa.

As the control center, the brain holds holographic maps of all the body's patterns, in diffuse and local forms (*opposite*), in the same way that your mind holds inner maps of your outer world. The relationship between the brain and the body is similar to that between the human mind and the cosmos, in that they both can contain a clear idea of the whole in one.

*Brain wave*s vary with mental phases. Alpha and beta waves hum during daily life, spaced-out delta waves visiting the young and sleeping. Fairy-tale theta waves imprint childhood experiences, and only resurface in daydreaming and very emotional adults.

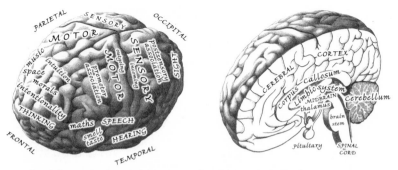

*Above left: some local cortical functions. Above right: the corpus callosum is a bundle of connections between the sides. The limbic system comprises peri-midbrain structures, like the hippocampus, and decides how significant or memorable you find things.*

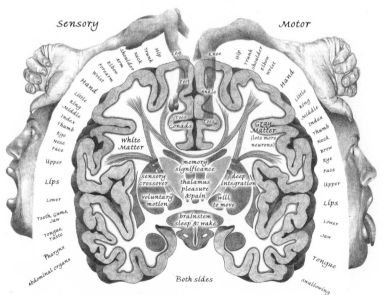

*Two slightly different mappings of the body held in gray matter*

# FIVE SENSES
## *and the twelve cranial nerves*

All but 12 nerves run to and from the brain in the spinal cord. Smell, I, relays directly to the cortex, which says "jasmine" or "?," while associated limbic circuits remember aromatic significance.

Retinal *rods and cones* turn off when light hits them, having been inhibiting bipolar cells, which flash the pattern of light and color down II. Two inverted images meet at the optic chiasm, split and half cross, then flip twice more before radiating onto the occipital cortices, which integrate to show you a single vision. The rest of your brain then decides what it means, and what to do about it.

Nerves III, IV, and VI swivel and focus the eyes; III also constricts the pupil, opposed by *sympathetic* nerves from ganglia in the chest.

Most chewing is driven by V the only cranial nerve to sense touch, from the dendrites in the skin of the face. Inner head spaces are under V, VII, and IX's intermingled sensorium, whose embryonic crossover explains tooth-ear confusion, and why odd things make you sneeze; VII also makes you drool, smile, and cry and tastes the sweet, sour, and salty front of the tongue. But the bitter back is IX's territory. The taste cortex warns the wanderer, *parasympathetic* X, of what may be heading the way of the innards.

Sound beating the eardrum transmits via 3 bones to the oval window of the 3-chamber *cochlear*, where cells are tensioned to resonate in tune. The spiral harpsong travels down VIII, along with motion data from the 3-D wing of the bony labyrinth *(opposite),* which senses both music and movement, so is crucial for dancing.

XII's fine-motor control of the tongue and larynx lets you speak, and XI (not shown) keeps your head up and shrugs your shoulders.

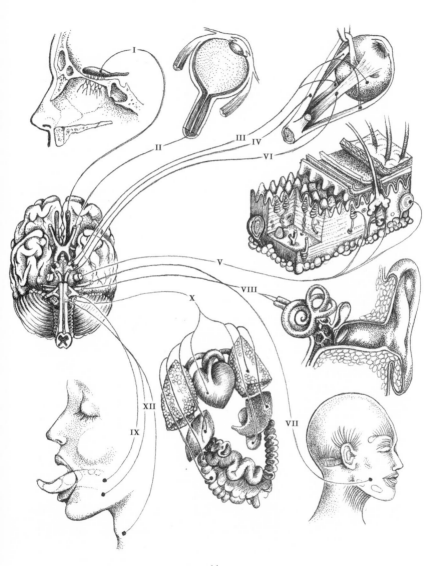

# MUSCLE AND BONE
## *a spring-loaded calcium dump*

What to do with heavy, double-charged, cell-trashing calcium? Dump most of it as shells, coral reefs, or endoskeletons in our case, and use the remaining few ions to tell it to hop and skip.

Calcium by nature alters the shape of molecules so is ideal for signal transmission, while energy-dependent processes involve phosphate transfer. In muscle cells, phosphates are preloaded onto ATPs between 2 huge proteins before they contract. After the decision to move, or look, or speak, every step in the long cascade from brain to muscle uses calcium triggers. Calcium finally dislodges these phosphates, and the spring-loaded muscle moves a bone, itself mostly made of calcium phosphate.

Your skeleton has 200 bones, plus 3 in each ear, where the tiniest muscle, stapedius, tensions the smallest bone, stapes.

Sensory endings in muscle called *proprioceptors* fire at a rate that depends on stretch, telling your brain the length of each muscle, so you can move accurately, even with your eyes shut. The loop delay in this sensory-brain-motor circuit causes the 10 Hz tremor you can see when you try to hold your hand still.

Muscles contract in groups and sequences, controlled by the motor cortex and cerebellum, where nerve cells fire in patterns to puppeteer complex routines, like writing or tumbling. These run more smoothly when not hampered by cortical overlay, also known as worry. Each cerebellar Purkinje cell has up to 100,000 dendrites receiving impulses from others, and this neural net can make new connections late into life to learn new skills, like flying.

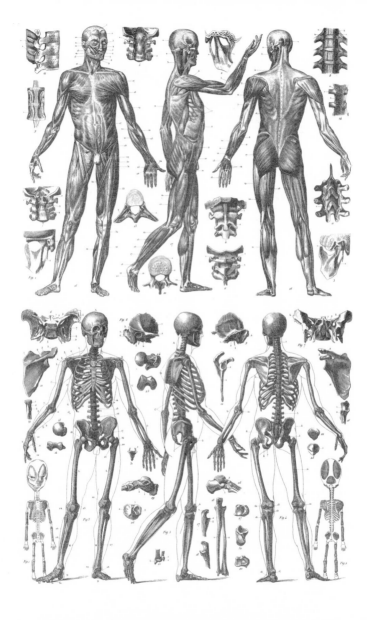

# ENDOCRINOLOGY
## *harmony of biorhythms*

Hormones dictate what phase body systems are in, from growth and *metabolism* to immunity and reproduction. They do this by switching influential processes on and off in distant cells. Seven major endocrine glands send hormones, via the blood, to all cells.

Any given hormone has certain "target" cells that respond to it. Some, like *thyroxin* and *insulin*, target most cell types, in type-variable ways. A few, like *growth hormone* and its opposite, *cortisol*, target all cells. But most endocrine business is self-regulation via control hormones, which only target cells in other glands.

The brain senses and governs the system in the hypothalamus, whose interplay with the pea-sized pituitary, on its stalk, releases control hormones to the lower glands, which make real hormones. Control loops time phases like the morning rush of cortisol, and the monthly surge and ebb of female *sex hormones*. Hormones start phases by switching on master genes that activate groups of other genes, mainly for enzymes that perform a related set of processes.

*Neuroendocrine* overlap is widespread, and it's hard to tell what's a hormone (see *paracrine*). The adrenal *medulla* is ectodermal (*p.18*), so is specialized nerve, which jibes with that *adrenaline* feeling.

Top of the pile, deep in the midbrain, is the mysterious pineal, primal neuroendocrine nexus. It times major phase changes, like night and day, seasons, puberty, menopause, and maybe death.

Indians have said for years that the simplest way to describe anyone's current state of health is as the activity and interplay of the 7 major chakras that sit up and down the midline.

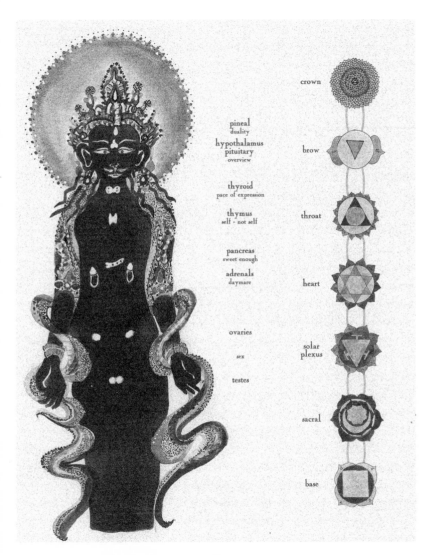

pineal
duality

hypothalamus
pituitary
overview

thyroid
pace of expression

thymus
self - not self

pancreas
sweet enough

adrenals
daymare

ovaries

sex

testes

crown

brow

throat

heart

solar
plexus

sacral

base

# HOMEOSTASIS
## *patterns in status quo*

Imagine all the various communications going on at this very moment between all the people on this planet, then cram all that into the same size as you, and that's about how busy it is in there.

This subconscious frenzy has one goal: to maintain equilibrium. Consider circulating calcium levels, honed by *parathyroid hormone* (PTH) and *calcitonin*. High calcium stimulates calcitonin release, to which signal bone cells respond by crystallizing calcium, while kidney cells excrete it, reducing current levels. PTH has opposite effects, and raises calcium levels when they fall. Together these 2 negative feedback loops keep calcium at the desired level.

Sets of similar loops balance everything from levels of ions to temperature, blood pressure, posture, vision, sanity, and so on, using chains of electrical and chemical signals. Signaling between cells creates integrated patterns of activity, by turning processes on and off inside cells using secondary intracellular signals. Gene switching, usually under hormonal control, tends to govern the longer-term variations in cellular activities, whereas short-term changes are mainly under electrical control by nerve impulses. Neuroendocrine integration of these regulatory feedback loops keeps the body's patterns in *status quo*, which is homeostasis.

Positive feedback loops are rare, as they amplify wildly. Apart from triggering *ovulation*, they occur in communication between humans, maintaining social equilibrium. They often involve the hormone *oxytocin*, which is busy during childbirth, breastfeeding, and orgasms, while oxytocin's limbic effects say "I love you."

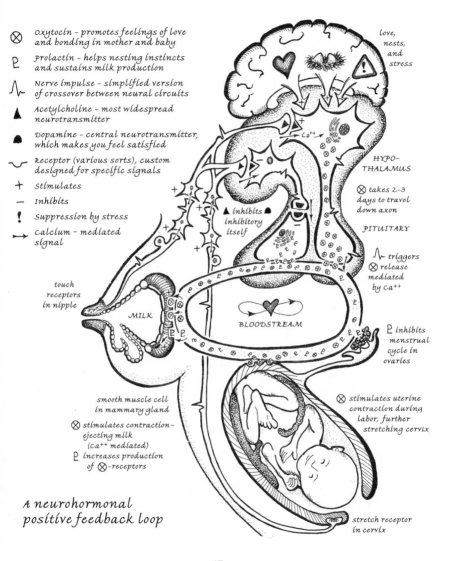

⊗ Oxytocin – promotes feelings of love and bonding in mother and baby

℘ Prolactin – helps nesting instincts and sustains milk production

⌇ Nerve impulse – simplified version of crossover between neural circuits

▲ Acetylcholine – most widespread neurotransmitter

● Dopamine – central neurotransmitter, which makes you feel satisfied

⌣ Receptor (various sorts), custom designed for specific signals

+ stimulates

− Inhibits

! Suppression by stress

⟶ Calcium – mediated signal

love, nests, and stress

HYPO-THALAMUS

PITUITARY

⊗ takes 2-3 days to travel down axon

⌇ triggers ⊗ release mediated by Ca++

▲ inhibits ● inhibitory itself

℘ inhibits menstrual cycle in ovaries

touch receptors in nipple

MILK

BLOODSTREAM

⊗ stimulates uterine contraction during labor, further stretching cervix

smooth muscle cell in mammary gland

⊗ stimulates contraction-ejecting milk (Ca++ mediated)

℘ increases production of ⊗-receptors

stretch receptor in cervix

A neurohormonal positive feedback loop

47

# NIGHT AND DAY
*which need each other*

The reason you feel as if you're falling apart after you've been up all night is because you are. The hormones that get you going in the morning, like the *catabolic* king, cortisol, make your body break down and use up its resources to go hunting and nesting. Maintenance and repair work, and *immunication*, are suppressed all day long, so that by bedtime, your body is literally disintegrating.

Growth hormone, the *anabolic* queen, rules by night, when your body reunifies the chaos left by the day. For this to happen, communication pathways that are closed by day open up, and like the intense repair of *inflammation*, the body's nightly healing is painful. You can feel where it's going to be busiest as you're dropping off, and but for sleep, the night would hurt.

Left unfettered, body processes are bursting to go, and have to be firmly inhibited if order is to prevail. Patterns of sensation particularly emerge through most nerves being damped. The day hormones' effects on sensory nerves are inhibitory, because all that sensation, while you're actually awake, would hurt too much.

The brainstem gets final short-term say over whether you're asleep, but longer-term patterns of sleep and wakefulness are under the control of the pineal, the original light-sensitive organ. This has evolved inward, but still receives light, via several neural routes, from the eyes. The neuroendocrine clock in the brain has an innate 25-hour rhythm, but is reset every day to Earth time, by the rising sun, via the pineal. So this gland, which oversees major life changes, is itself ruled by our motion in the solar system.

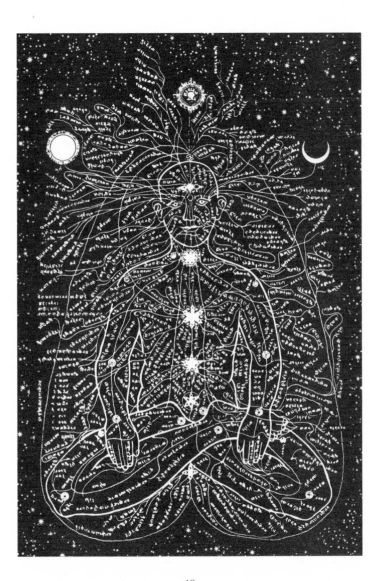

# WHAT IS IT FOR?
## *and where is it going?*

The Earth has been revolving around our Sun for 5 billion years or more, and for most of that time life has been coevolving and refining itself to produce beautiful and complex life-forms like the human body, which has appeared in the most recent twinkling of the cosmic eye. The *entelechy* with which evolution realizes new possibilities resembles less a series of lucky breaks than a system that has the idea first, then works out a way to make it happen.

The biggest recent changes in human form are jaw shape and teeth alignment, which have evolved in the last few thousand years to cope with cooked food. Other subtler progress may be taking place under our noses, but our use of technology is racing ahead of any longer-term refinements in bodily structure, which has changed little since the image opposite was painted on a cave wall in Spain some 8,000 years ago. Unlike our foraging ancestors, and all other organisms as far as we know, we are in the position of having minutely dissected and examined our own insides, and we can influence them in hitherto unthinkable ways. Now that we know what all the little bits do, we may find some of the finer determinants of human life in the study of overall field effects, both of substances and electromagnetic fluxes in and around the body.

Meanwhile, every moment of your life your body is hectically sending signals between its various parts, as electrochemical patterns, in order to enable you to do things like read this book. Perhaps this busy inner messaging hints at what it's really for, what humans are getting better and better at, which is communication.

# SMALL GLOSSARY OF ITALICIZED TERMS

**ADRENALINE** 44, 56; fight, might, or flight monoamine neurohormone.

**AMINO ACID** 10, 55, 56; *protein* building blocks, consisting of H₂N(amine)-C-COOH (acid). A side chain on the middle carbon defines the twenty types used in humans.

**AMOEBA** 32; morphing unicellular blood blob.

**ANABOLISM** 48, 55, 56; rebuilding, storage, and integrative *metabolism* in general.

**ANTIBODIES** 34; immunoglobulins that often need complementary activation.

**ANTIGEN** 34; anything that excites immune reaction to it; usually a microbial protein.

**AQUEOUS HUMOR** 2; specialized extracellular fluid in the anterior chamber of the eye.

**ARCHAEBACTERIA** 12; earliest organisms.

**ATP** 14, 30, 42, 54, 55; adenosine triphosphate.

**ATRIO-VENTRICULAR VALVES** 28; tricuspid (right), and bicuspid (left, from the body's point of view; the inner heart valves.

**ATRIA** 28; the two upper chambers of the heart.

**AUTONOMIC SYSTEM** 16, 36; subconscious motor control of smooth muscles (in tubes like gut, bronchioles, and arterioles), heart muscle, and glands; see (*para-)sympathetic*.

**AXON** 36; delivers a nerve impulse from a *neurosome* to another cell.

**AXON HILLOCKS** 36; area of a nerve cell between *neurosome* and *axon*; outgoing impulses start here, when the sum of local voltages reaches the firing threshold.

**BACTERIA** iv, 10, 12, 35, 55; genetically unified unicellular organisms with no nucleus.

**BARR BODY** 8; humans run on one X-chromosome, so females, who have 2 Xs, inactivate one in each cell by scrunching it into a Barr body; half maternal and half paternal. Xs are inactivated while an embryo, giving "mosaic" effects in females, and is why identical twin girls are less alike than boys. In the fetal part of the placenta, and in all marsupial cells, Barr bodies are all paternal.

**BASES** 6, 10; coding nucleotides binding as pairs across *DNA's* double helix; adenine (*a*) pairs with thymine (*t*), and cytosine (*c*) with guanine (*g*); RNA replaces *t* with uracil (*u*).

**BLOOD** 6, 16, 18, 20, 24, 26, 28, 30, 32, 44, 46, 54, 55, 58

**BLOOD FLOW** 28, 56; tissue-variable flow rates, and pooling effects, confuse the issue.

**BONE** 16, 18, 20, 32, 42, 46, 58.

**BRAIN** 14, 17, 19, 20, 26, 28, 36, 38, 40, 42, 44, 47, 48.

**BRAINSTEM** 26, 38, 48; the oldest part of your brain, comprising mainly pons and medulla.

**BRAIN WAVES** 38; beta (alert, 15-40 Hz), alpha (relaxed, 8-14 Hz), theta (daydream, 4-8 Hz), and delta (asleep, 1-4 Hz) waves.

**BRONCHI** 26; larger tubes of your lungs.

**CALCITONIN** 46, 56; from the thyroid; one of two main calcium-regulating hormones.

**CAPILLARIES** 24, 26, 30, 32, 54, 56; blood tissues exchange interface of vast area.

**CATABOLISM** 48, 55, 56; "use it up" and "wear it out" aspect of *metabolism*.

**CELL** 8, 10, 12, 14, 16, 18, 20, 22, 24, 26, 28, 30, 32, 34, 36, 40, 42, 44, 46, 54, 55, 56.

**CHAKRAS** 4, 44; Sanskrit for "wheels"; interfaces between physical and "etheric" bodies, visible to some individuals.

**CHANNELS** 14, 36, 54, 56; transmembrane structures, allowing specific ions or molecules to pass through.

**CHI** 4; Chinese for universal life force.

**CHROMOSOMES** 8, 10, 16; long strands of DNA, linear in all but mitochondria, which have one small circular chromosome; the mom-dad pairs in humans; carry *genes*.

**COCHLEAR** 40; spiral center of hearing.

**CODONS** 10; 3-base units of RNA/DNA that specify either an amino acid, or start or stop.

**COMPLEXION** 6, 58; mix of constitution and temperament, was more than skin deep.

**CONCEPTUS** 8; the fertilized ovum.

**CORTEX** 38, 40, 42, 44, 56; outer part or layer of an organ, e.g., brain, kidney, or adrenal.

**CORTISOL** 44, 48, 56; daytime adrenal hormone; levels surge as you are waking up.

**DENDRITES** 36, 42; receptive tendrils of nerve cells; converse of *axons*.

**DESCARTES** 44; French thinker; 1596-1650.

**DNA** 6, 8, 10; deoxyribonucleic acid; uncoiled, your total DNA would stretch to the moon and back several times.

**ECTODERM** 18, 44, 57; outer germ layer.

**ENDOCRINOLOGY** 6; study of hormones.

**ENDODERM** 18, 57; inner lining germ layer.

**ENDOSYMBIOSIS** 12; house-sharing.

**ENTELECHY** 50; the way life realizes its patterns with an apparent end in sight.

**ENZYMES** 14, 24, 44, 54; proteins that act as catalysts; usually working with coenzymes, they potentiate molecular breakdown, synthesis, and transformation.

**EUKARYOTIC** 12; complex cell with a nucleus.

**EXTRACELLULAR FLUID** 14, 16, 30, 32, 54, 57.

**FACTORS** 44, 56; in endocrinology, control hormones, which stimulate or inhibit the synthesis and release of other hormones, thus regulating their levels.

**GAMETE** 8; *haploid* intergenerational cells.

**GANGLIA** 36; nodal colonies of nerve cells in the brain and spinal cord, or just outside the spine (which has 2 *sympathetic* and 2 *sensory* ganglia per segment).

**GATING** 4; central nervous system blocking effect by a pattern of peripheral stimuli.

**GENE** 8, 10, 12, 14, 16, 18, 22, 56; codes for a protein (or RNA unit); humans have about 30,000 genes but can make 10 times as many proteins, as RNA combines recipes.

**GENOME** 10; total DNA for an organism.

**GLIAL CELLS** 36; metabolic and structural support for nerves in brain and spinal cord.

**GLOMERULI** 26; initial filtration subunits of the kidney; each consists of a high-pressure capillary bundle and collecting capsule.

**GONADS** 18, 21; ovaries or testes.

**GRANULOCYTES** 34; white blood cells, mostly neutrophils that deal with bacteria.

**GROWTH HORMONE** 44, 48, 56; pituitary *peptide*, rules regeneration and integration.

**HAPLOID** 9; single set of chromosomes.

**HEMOGLOBIN** 26, 30; gas-carrying molecule.

**HOMO** 38; Latin for "man," as in *H. sapiens*.

**HORMONES** 6, 18, 20, 44, 46, 48, 56; the most influential signaling molecules in your body.

**HYDROGEN BONDS** 10; loose proton links between molecules, especially important in life-temperature aqueous systems (like you).

**IMMUNICATION** 48, 55; the many ways in which immune-system cells communicate.

**INFLAMMATION** 48; it hurts because tissues are being opened up to allow repair.

**INHIBIT** 36; restrain, dampen, and slow processes.

**INSULIN** 44, 56; anabolic pancreatic hormone.

**INTRACELLULAR FLUID** 15, 16, 54; cytosol.

**INTRONS** 11; palindromic or unwanted regions of mRNA that are snipped out before protein synthesis.

**IONS** 14, 36, 42, 46, 54; charged atom(s), e.g., Na⁺, Cl⁻, PO₄³⁻, usually in water.

**IRISMEISTERS** 2; holographic iris wizards.

52

**KIDNEYS** 16, 18, 20, 26, 46, 58.

**KUNDALINI** 4; awakened Earth energy.

**LIMBIC SYSTEM** 38; around the border (*limbus*) of the midbrain; comprises olfactory tracts (smell), areas of cortex such as the hippocampus, parts of the thalamus and hypothalamus, and other esoterica.

**LIVER** 17, 20, 24, 30, 58.

**LUNG** 17, 20, 26, 28, 30, 32, 58.

**LYMPHOCYTES** 32, 34; white blood cells; many types performing cooperative functions in the immune system.

**MACROPHAGES** 30, 32, 34; white blood cells that engulf and digest unwanted stuff.

**MARROW** 30, 32, 58; highly vascular tissue of mesodermal origin in middle of bones, where stem cells give rise to all types of blood cells.

**MEDULLA** 44, 56; middle of an organ.

**MEIOSIS** 8; specialized cell division in *gonads* to form *gametes*.

**MEMBRANE** 11, 14, 36, 57; the cell wall; see *plasma membrane*.

**MESODERM** 18, 57; middle germ layer, makes blood, solid organs, muscle, bone, and marrow.

**METABOLISM** 44, 55, 57; the daily molecular transformations your body makes, divided broadly into *anabolism* and *catabolism*.

**MHC PROTEINS** 34; major histocompatibility complex; on cell walls; recognized as self.

**MITOCHONDRIA** iv, 11, 12, 14, 26, 30, 54, 55, 56; tiny red modified *archaebacteria*, which outnumber your cells by hundreds to one; electroenergetic grubs, they feed on oxygen and carbon, keep you alive and warm, and make a glass of fresh water a day.

**MITOSIS** 8, 14; standard cell division.

**MOLECULE** 6, 10, 14, 16, 30, 42, 54, 55; bonded structure made of atoms.

**MOTOR NERVES** 36; tell muscles to contract.

**MUSCLE** 2, 16, 19, 21, 28, 36.

**MYELIN** 36; insulating nerve sheath made by Schwann cells; speeds up nerve conduction, as the impulse hops along inside the cell.

**NA-K-ATP PUMP** 14, 54; thousands per cell wall, generate electricity across *membrane*.

**NERVE** 14, 16, 19, 20, 36, 38, 40, 42, 47, 48, 50

**NEUROENDOCRINE SYSTEM** 44, 48, 57; brain, nerves, and hormones, as a totality that controls the body.

**NEUROMUSCULAR JUNCTION** 56; synapse between *axon* and muscle cell, where acetylcholine is usually the neurotransmitter.

**NEURON** 36; another word for nerve cell.

**NEUROSOME** 36, 38; main torso of a nerve cell, as opposed to its *axons* and *dendrites*.

**NEUROTRANSMITTER** 36, 56; molecule that perpetuates an electrical signal between cells.

**NUCLEUS** 10, 12 , 30, 56; brain of a cell, home of your DNA, enclosed by a selectively porous nuclear membrane; each nucleus houses all your genes, mostly in the dense central nucleolus, supercoiled around histone proteins, while the few genes in use uncoil to be read by mRNA in the nuclear *cortex*.

**OVULATION** 46; release of an *ovum* or two.

**OVUM** 8, 52; female *gamete*, fattest cell of all.

**OXYTOCIN** 46, 56; posterior pituitary nonapeptide hormone of love, forgetting, and letting go; sister to *vasopressin*.

**PACEMAKER CELLS** 28; all cardiac cells fire rhythmically automatically; the fastest ones, pacemaker cells, spark the electromuscular wave over the heart from the right *atrium*.

**PARACELSUS** 6; influential healer and natural mystic; c. 1491-1543.

**PARACRINE** 44, 55; type of communication between cells using long-chain fatty acids; e.g., prostaglandins, thromboxanes, interleukins; in tissues, locally in blood, and long distance.

**PARASYMPATHETIC**; division of *autonomic* nervous system, running digestive, anabolic, and low-alert relaxational states e.g., orgasms.

**PARATHYROID HORMONE** 46, 56; releases calcium from bone and slows its excretion.

**PEPTIDE** 54, 55, 56; protein (usually smallish), also the bond between 2 *amino acids*.

**PLACENTA** 18, 47, 56; mother-fetus interface.

**PLASMA** 16; fluid (noncellular) part of blood.

**PLASMA MEMBRANE** 14; semifluid, self-forming wobbly phospholipid bilayer.

**PLATELETS** 32; cell fragments involved in clotting; signal using modified fatty acids.

**PRANA** 4; Indian *chi*.

**PROPRIOCEPTORS** 42; stretch-sensitive dendritic nerve endings in skeletal muscle.

**PROTEINS** 10, 14, 24, 30, 32, 34, 42, 55; chains of *amino acids*, connected by *peptide* bonds, which fold into more complex 3-D forms via *hydrogen bonds* and disulphide bridges.

**PULSE** 28; in arteries, the shock wave from cardiac contraction, not the blood flow itself, so carries a mix of heart and vessel character.

**PUMPS** 14, 54; static *enzyme* chains that move electrons, protons, ions, or molecules through *membranes*, or between molecules.

**QANUN** 26; a systematic compendium of current knowledge (here by Ibn al-Nafis).

**RECEPTORS** 14, 56; in cell wall or cytosol;

change shape to perpetuate a signal from specific hormones and neurotransmitters.

**RNA** 10; ribonucleic acid, older than DNA, less durable, but less spliceable; single-stranded in humans; (*see bases*).

**RODS AND CONES** 40; retinal cells that sense light (rods) and color (cones).

**SEMILUNAR VALVES** 28; usually 3-leaved; stop blood refluxing into *ventricles*.

**SENSORY NERVES** 36, 40, 42, 48; their long *dendrites* form specialized endings to detect particular stimuli e.g., touch, light, heat, pressure, vibration, cold, taste, smell, etc.

**SEX HORMONES** 44, 56; estrogens and progestogens (mainly women), and androgens like testosterone (mainly men).

**SHIVA AND SHAKTI** 4; divine Indian couple, manifesting *yin* and *yang* principles.

**SPLEEN** 21, 25, 30, 33.

**SRY-1** 18; initial gene in male cascade.

**STATUS QUO** 46; dynamic equilibrium of existing conditions, current state of play.

**SUPERCOILING** 10; keep twisting a rubber band to make a simple superhelix.

**SYMPATHETIC**; division of *autonomic* nervous system, running high-alert catabolic phases of action e.g., hunting and lovemaking.

**SYNAPSE** 36; joint between 2 nerve cells; a neuromuscular junction is a sort of synapse.

**TISSUE** 2, 14, 16, 18, 20, 22, 30, 32, 55.

**THYMUS** 34, 45, 56; immune endocrine gland.

**THYROXINS** 44, 56; hormones secreted by the thyroid; major metabolic influence.

**UTERINE MILK** 18; nutritious early food.

**VASOPRESSIN** 56; blood pressure, memory, and salt-balance hormone.

**VENTRICLES** 28; 2 lower chambers of the heart; pump blood out to body and lungs.

**VILLI** 24, 26; intestinal fingers, with microvilli, central capillaries, and a lymphatic, which absorb all your food from the small intestine.

**VIRUSES** 10; parasitic polyhedral replicators.

**VOLTAGE GATE** 36; door of a *channel* that opens or shuts depending on local voltage.

**VOLTAGE GRADIENT** 36; change in electrical charge per unit distance.

**WHITE MATTER** 38; inner parts of brain and outer parts of spinal cord; mostly consists of axons, getting its color from their myelin.

**XII** 40; I olfactory, II optic, III oculomotor, IV trochlear, V trigeminal, VI abducent, VII facial, VIII vestibulo-cochlear, IX glosso-pharyngeal, X vagus, XI accessory, XII hypoglossal.

**YANG** 4, 58; positive, male, light, active.

**YIN** 4, 58; negative, female, dark, receptive.

**YOLK SAC** 18; embryonic food and cell store.

Head  Body  Tail  End piece

# WATER, MINERALS, AND FLUID COMPARTMENTS

When biologists are asked what sort of life we have on Earth, the reply used to be "carbon based," but now it is "water based." Water's unique properties allow all the biochemical and electrical transformations in life to take place, and it is the main constituent of your body. Body fluids come in three major compartments: blood, *extracellular fluid* (ECF, the salty soup between cells), and *intracellular fluid* (ICF, the viscous cytosol inside cells). Nearly half of blood volume is cells, mainly red cells, and a percent or two of white cells and platelets. The fluid portion of blood, *plasma*, is similar to extracellular fluid, but very large proteins like albumin stay in the blood, contributing to its osmotic ability to keep water within capillaries. About 30 liters of fluid seeps out of capillaries per day, and all but 3 liters (drained as lymph) seeps back in, carrying cell products.

Fluid ratios vary with body type, gender, and age, but an adult has, in liters, about 5 in blood, 9 in ECF, and 29 in ICF.

Most blood is pooled as a slow flowing reservoir in veins, between carrying everything else around. Brain and kidneys take a fifth of (nonlung) flow each, and the busiest-current organs take most of the rest, especially muscles (if you're running), gut and liver (after a meal), or skin (if you're hot).

In terms of dissolved molecules (rather than ions), ICF composition varies hugely, depending on cell type and phase of activities. Plasma carries all circulating substances, and ECF mediates between ICF and plasma.

Charged ions are kept at stable levels in fluid compartments, because life processes operate within narrow local electrical and acid-base windows. An ion is surrounded by a multilayered morphing "aquahedron" (*i.*) of water molecules, defining its sphere of influence. Potassium and sodium ions both carry a single positive charge, but though $K^+$ is heavier, it is less electrodense, so has a smaller, less influential aquahedron. This makes it the tame cation of choice for intracellular use and for stabilizing the membrane voltage by seeping through small cation channels. Sodium performs

the faster, more electrodense processes like rushing through voltage-gated channels and is ejected from cells by Na-K-ATP pumps, 3 $Na^+$ out to 2 $K^+$ in per ATP molecule. The salty sea of ECF has chloride ($Cl^-$) as its main anion, while phosphates and proteins carry most ICF negative charges.

## MAJOR ION CONCENTRATIONS (mmolL⁻¹)

|  | Plasma | ECF | ICF |
|---|---|---|---|
| *Sodium* | 144 | 146 | 11 |
| *Potassium* | 4 | 4 | 148 |
| *Calcium* | 4 | 3 | 0.1-3 |
| *Magnesium* | 2 | 2 | 35 |
| *Chloride* | 104 | 118 | 4 |
| *Bicarbonate* | 25 | 27 | 13 |
| *Phosphate* | 2 | 2 | 40-100 |
| *Protein anions* | 20 | 1 | 52 |
| *Others* | 5 | 6 | 60 |

Of other minerals, iron (mainly for hemoglobin) is more abundant in men, whereas copper is higher in women. Rarer elements (Cu, Mg, Mn, Zn, Cr, Se, etc.) are mainly enzyme components, for instance molybdenum, the rarest element you need, which is in xanthine oxidase. A cobalt sits in the middle of vitamin $B_{12}$ (*ii.*), a coenzyme, like most vitamins.

Sulfur appears mainly in proteins, playing a special role in forming their 3-D shape. Iodine is used in thyroxine synthesis, being super added to iodine-containing peptides.

A fourth fluid category is special extracellular secretions, which includes the cerebrospinal fluid the brain floats in, the eyes' aqueous and vitreous humors, and the ears' endolymph and perilymph, and also any fluid or mucus matrix that cells deposit around them for lubrication, insulation, protection, and as an extended molecular territory for cellular activities.

About nine-tenths of your water comes in what you eat and drink; the rest is made as fresh, new water molecules, a little during protein synthesis, but most of it during mitochondrial ATP synthesis.

54

# MOLECULAR NUTRITION

What you eat and drink is water and minerals, lipids (fats), carbohydrates (sugars), proteins, and vitamins. The complex molecules in your food have already been synthesized by the combined efforts of bacteria, fungi, plants, and animals (if you eat meat). The energy source for the synthesis of complex organic molecules is the Sun.

Your body can interconvert most substances, but some molecules the body cannot synthesize are essential nutrients, mainly certain amino acids, lipids, and vitamins. Everything else can be made from other molecules.

*Catabolism* is the breakdown of complex molecules into simple ones that are used to supply energy or to build other compounds. *Anabolism* is the synthesis of complex compounds. These processes require energy input, which in most nonmitochondrial pathways comes as *ATP* (a.).

*a. adenosine-triphosphate*

The *Krebs cycle* is the multistep circular pathway by which mitochondria produce ATP. The brain has to use *glucose* (b.) to fuel this process, and other cells prefer to, as it can be split via a simple eleven-step process into two *pyruvates* (e.g., c.), which enter the Krebs cycle two steps later. Most of the pathways of sugar, fat, and protein catabolism converge around here. Short-term lack of oxygen diverts pyruvate to *lactate* (d.) in an "emergency" pathway.

*b.*

*c.*

*d.*

Dietary sugars can be simple like glucose, or the fruit sugar *fructose* (e.), disaccharides like *sucrose* (b. + e.) or *lactose* (f.), the milk sugar that is the main nonplant carbohydrate you're likely to ingest, or complex *sugar polymers* (starches). Bean sugars are initially digested by gas-producing bacteria, hence the flatulence. Sugars can be stored as *glycogen*, a glucose

*e.*

*f.*

polymer, synthesized by liver and skeletal muscle cells as a rapid access reserve, but 99% of energy storage is as fat, mainly *triglycerides* (e.g., g.), the destination of all excess sugars,

*g.*

proteins, and lipids. The two essential dietary fats are *linoleic* and *linolenic acid* (h.), found in all plant cells. Other dietary fats

*h.*

are *cholesterol* (essential in cell walls and for making steroid hormones, bile salts, and vitamin D), and *phospholipids* (like lecithin), the main constituents of cell walls. Derivatives of the 20-carbon arachidonic acid, made from linolenic acid, are also used as *paracrine* messengers between cells, both within tissues and in movable operations like damage repair and *immunication*.

*j.   R = radical*

Plants make all amino acids (j.), and we can synthesize 10 of the basic 20, the rest being essential amino acids like *tryptophan* (k.). *Peptide* bonds join amino acids to make proteins of all sizes, from *dipeptides* (two amino acids) to huge muscle proteins like *titin*, a chain of 27,000. If you have to use proteins for energy, you remove their nitrogen-containing parts, creating ammonia, which is also made at many other metabolic stages. Birds

*k.*

excrete excess nitrogen as urates in semisolid guano, for mammals excrete ammonia dissolved as *urea* (l.), which is the main reason you have to drink so much water. By the time you eat your food, the hard work of synthesis has already been done. Our metabolic processes have evolved with the organisms we eat, and use the same substances in different pathways, so a diet of good water, fruit, nuts, roots, vegetables, and some extra protein, especially in childhood, will supply all your needs.

*l.*

# NEUROTRANSMITTERS

Nerves send their messages to other nerves or muscles using neurotransmitters (NTs). NTs tend to be small molecules, mainly amino acid derivatives, which travel across the gaps in synapses or neuromuscular junctions, binding to receptors on target cells, which opens ion channels ($Na^+$ and/or $Ca^{2+}$ in excitatory neurons, and $K^+$ or $Cl^-$ in inhibitory ones). Receptors come in several subtypes for most NTs, having variable or polar effects in different cells, and are often named after natural drugs that excite or block them, like nicotinic and muscarinic receptors for ACETYLCHOLINE (ACh). ACh is the purest NT; all the others have some hormonelike phasic activity on certain cells, like NORADRENALINE (NA). ACh and NA are the main NTs of the peripheral nervous system and can be excitatory or inhibitory and receptor-dependent. ACh and NA are also central-nervous system NTs. Other major CNS NTs are as follows:

SEROTONIN (5-HT): tryptophan derivative, gut neurohormone, and rare but crucial midbrain/brainstem NT. Mood, humor, significance of perceptions on all levels.

DOPAMINE: unlike 5-HT, mainly excitatory, midbrain motor, and limbic motivation/reward/satisfaction areas.

HISTAMINE: inhibitory in hypothalamic temperature and arousal circuits. Common peripheral inflammatory signaler.

GABA: prevalent local inhibitor in brain. Relaxation in general, as after a meal. Many addictive drugs mimic it.

GLYCINE: spinal cord GABA, blocked by strychnine.

GLUTAMATE and ASPARTATE: excite nonlocal integration.

ENDORPHINS and other PEPTIDE OPIOIDS: inhibitory, widespread, especially pain tracts. Transitory blissful peace.

SUBSTANCE P: excites, "antiendorphin," chili, P for pain.

NITRIC OXIDE: blood flow, and non-CNS local control of heart/artery/gut muscles. In nerves to penis and clitoris.

# MAJOR HORMONES BY GLAND

"Hormone" is a loose term these days, with new messaging systems being discovered regularly. "Proper" hormones are either amino acid derivatives (as amines, polypeptides, or glycoproteins) or cholesterol derivatives (called steroids, made by the adrenal cortex, testis/ovary, and placenta).

PINEAL: releases many newly discovered peptides, but its only even partially understood product is *melatonin*, made from *serotonin* during darkness. Dual rhythms (pp.44,48).

HYPOTHALAMUS: secretes only release and inhibitory *factors* controlling pituitary, via dedicated capillary loops.

ANTERIOR PITUITARY: controls lower glands with trophic hormones, like *ACTH* (adrenal cortex), *TSH* (thyroid), and *FSH/LH* (ovaries and testes). Controls pigmentation (*MSH*), growth (*GH, p.48*), fatness (*LPH*), and milk production (*prolactin, p.46*).

POSTERIOR PITUITARY: makes 2 nonapeptide hormones (9 amino acids) that differ by just 2 amino acids: *oxytocin*; (letting go of feelings, fetus, milk, and memories) and *vasopressin*; (holding on to water, memories, and emotions).

THYROID: speeed of development, metabolism, growth, and energy. *Thyroxines* are iodine-added peptides that travel into cells and either bind with a cell's *zinc-finger proteins* (off to the nucleus to regulate transcription rates of sets of master genes), or short-circuit mitochondrial ATP production, resulting in release of energy as heat.

PARATHYROID: *PTH* opposes thyroid's *calcitonin* (p.46).

THYMUS: childhood immune development (p.34).

PANCREAS: islet cells secrete a) *glucagon* (29 amino-acid peptide), catabolic king of gut hormones, which opposes b) *insulin* (two-unit peptide, 21+30 amino acids), which stimulates storage, repair, and anabolism in general. Other pancreatic hormones control digestive signals and processes.

ADRENAL MEDULLA: neurohormones *adrenaline* and *NA*.

STEROID HORMONES: like thyroxines, are small, and bind other zinc-finger proteins, with pervasive effects, and also act via cell-wall receptors, like most other hormones.

ADRENAL CORTEX: *catabolic* steroids (e.g., *cortisol*) inhibit gene expression, others rule salt-water balance. Makes extra sex steroids, male and female, in both sexes.

OVARIES: womanhood via *estrogens* and *progesterones*.

TESTES: manhood via androgens like *testosterones*.

PLACENTA: *steroids* and enormous *pregnancy peptides*.

ALL OTHER ORGANS: make hormonelike substances.

# THREE-FOLD SYSTEMS

| | ECTODERM | MESODERM | ENDODERM |
|---|---|---|---|
| *Embryonic germ-cell layer* | | | |
| *Body type* | thin, wiry, active | stocky, strong, cyclic | rounded, steady, slow |
| *Ayurvedic dosha & humor* | VATA - WIND | PITTA - BILE | KAPHA - PHLEGM |
| *Tibetan elements* | AIR & SPACE | FIRE | EARTH & WATER |
| *Arabian-European alchemy* | mercury/communication | sulfur/energy | sal/structure |
| *Fluid compartment & cation* | extracellular/calcium | blood/sodium | intracellular/potassium |
| *Body systems* | neuroendocrine, integration | metabolic, immune | assimilation, excretion |
| *Tissue & organ color* | brain, nerve, skin, white/brown | muscle, liver, heart, red | gut, lung, tubes, transparent |
| *Teeth* | long, crooked, brown-black | medium, hard, gray-yellow | strong, large, white-blue |
| *Eyes* | beady, black holes | penetrating, yellow-red | big, open, clear |
| *Psyche* | thought, integration, desire | compassion, vision, anger | acceptance, stability, greed |
| *Life* | inspirational, myriad, future | phasic, powerful, present | devoted, belonging, past |
| *Memory* | quick/recent | all-encompassing | slow, long, deep |
| *Sleep & dreams* | shallow, mountains, flying | refreshing, action, light | sound, water, clouds |
| *Trouble* | cold, falling, isolation | heat, violence, gloom | drowning, stasis, loss |

# WESTERN FOUR-FOLD SYSTEM

| *Type* | ARTISAN | GUARDIAN | IDEALIST | RATIONAL | *Plato* |
|---|---|---|---|---|---|
| *Element* | FIRE | EARTH | AIR | WATER | *Nature* |
| *Qualities* | hot & dry | cold & dry | hot & wet | cold & wet | *Weather* |
| *Solid* | tetrahedron | cube | octahedron | icosahedron | *Geometry* |
| *Humor* | SANGUINE | MELANCHOLIC | CHOLERIC | PHLEGMATIC | *Galen* |
| *Fluid & organ* | blood & liver | bile & intestines | plasma & lungs | mucus & kidneys | *Medieval* |
| *Extremity* | arms | legs | head | genitals | *Body politic* |
| *Systems* | electric-metabolic | structural-permanent | breath-integration | reproduction-homeostasis | *20th C.* |
| *Temperament* | changeable | industrious | inspired | curious | *Paracelsus* |
| *Personality* | influential | conscientious | dominant | steady | *Haines* |
| *Jungian types* | ISFP, ISTP, | ISFJ, ISTJ, | INFJ, INFP, | INTJ, INTP, | *Jung* |
| *(see below)* | ESFP, ESTP | ESFJ, ESTJ | ENFJ, ENFP | ENTJ, ENTP | |
| *Nature* | exploitative | hoarding | receptive | marketing | *Fromm* |
| *Mindedness* | probing | scheduling | friendly | tough | *Myers-Briggs* |
| *Self-Image* | artistic, audacious | dependable, beneficent | empathic, benevolent | ingenious, autonomous | *Keirsey* |
| | adaptable, hedonistic | respectable, stoical | authentic, altruistic | resolute, pragmatic | |
| *Orientation* | optimistic, cynical | pessimistic, fatalistic | credulous, mystical | skeptical, relativistic | |
| | here, now | gateways, yesterday | pathways, tomorrow | intersections, intervals | |

*Jung's categories are based on four axes: Introverted or Extroverted (I or E), Intuitive or Sensing (N or S),*
*Feeling or Thinking (F or T), and Perceiving or Judging (P or J). See, too, page 7.*

# CHINESE SYSTEM

| | 木 WOOD | 火 FIRE | 土 EARTH | 金 METAL | 水 WATER |
|---|---|---|---|---|---|
| *Yin Organ (zang)* | liver | heart & pericardium | spleen | lung | kidney |
| *In charge of* | strategy, plans | pulse & protection | processing | qi of heaven | water control |
| *Muscle meridians* | leg jue yin | arm shao & jue yin | leg tai yin | arm tai yin | leg shao yin |
| *Yang Organ (fu)* | gallbladder | sm. intest & trip heat | stomach | large intestine | bladder |
| *In charge of* | decisions | digestion & regulation | fermentation | elimination | storing liquids |
| *Muscle meridians* | leg shao yang | arm tai & shao yang | leg yang ming | arm yang ming | leg tai yang |
| *Tissue* | tendons, sinews | blood vessels | flesh, muscles | skin, pores | bones, marrow |
| *Reflects in* | nails | face, complexion | lips | body hair | head hair, teeth |
| *Orifice, sense* | eyes, sight | tongue, speech | mouth, taste | nose, smell | ears, hearing |
| *Body fluid* | tears | sweat | drool | mucus | spittle |
| *Taste, odor* | sour, rancid | bitter, scorched | sweet, aromatic | pungent, rotten | salty, putrid |
| *Voice, emotion* | shouting, anger | laughing, elation | singing, worry | crying, sadness | groaning, fear |
| *Aspects of Shen* | *hun*, soul life | *shen*, awareness | *yi*, memory, intellect | *po*, instinct | *zhi*, willpower |
| *Stage* | germination | growth | transformation | harvest | storage |
| *Season* | spring | summer | late summer | autumn | winter |
| *Injurious climate* | windy | hot | damp, humid | dry | cold |
| *Direction* | east | south | center, middle | west | north |
| *Color, time* | green, dawn | red, midday | yellow, afternoon | white, dusk | black, midnight |
| *Animals* | dragon, sheep | phoenix, fowl | pangu, snake, ox | tiger, dog | tortoise, pig |
| *Planet* | Jupiter ♃ | Mars ♂ | Saturn ♄ | Venus ♀ | Mercury ☿ |
| *Food* | wheat, lemons | pepper, alfalfa, greens | millet, potato, fruit | rice, ginger, air | beans, seafoods |
| *Yin-yang* | lesser yang | utmost yang | center | lesser yin | utmost yin |
| *Mode, note* | *jiao*, e | *zhi*, g | *gong*, c | *shang*, d | *yu*, a |

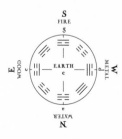